TO THE PARENT

What an accomplishment it is to count to 100, and
what an eye opener it will be when your child realizes
that there are many different ways to get there! This book
is a fun introduction to the idea that numbers can be
grouped into units—like fives, tens and twenty-fives.
Learning to manipulate numbers is important groundwork
in improving math skills. Have fun!

100 Days of School

By Trudy Harris Art by Beth Griffis Johnson

MILLBROOK PRESS / MINNEAPOLIS

To Jay
T. H.

With love to my husband,
Nolden. And . . . in memory of
a wonderfully gifted instructor
and friend, Dwight Harmon.
B. G. H.

Text copyright © 1999 by Trudy Harris
Illustrations copyright © 1999 by Beth Griffis Johnson

Millbrook Press
A division of Lerner Publishing Group, Inc.
241 First Avenue North
Minneapolis, MN 55401 USA

Website address: www.lernerbooks.com

Library of Congress Cataloging-in-Publications Data
Harris, Trudy.
100 days of school / Trudy Harris: illustrations by Beth Griffis Johnson.
p. cm.
Summary: A series of rhymes illustrates different ways to count to 100 such as by adding the
ten toes of ten children or ninety-nine train cars plus one caboose.
ISBN: 978–0–7613–1271–0 (lib. bdg. : alk. paper)
1. Addition—Juvenile literature. 2. Hundred (The number)—Juvenile literature. [1. Addition. 2.
Hundred (The number) 3. Counting.] I. Johnson, Beth Griffis, 1967– ill. II. Title.
QA115.H38   1999
513.2'11[E]—dc21   98-18952

Manufactured in the United States of America
13 – DP – 5/1/12

If you go to school for 95 days, and then go 5 more days, what do you get?

Smarter and smarter.

And . . .

(how cool)
100 DAYS OF SCHOOL!

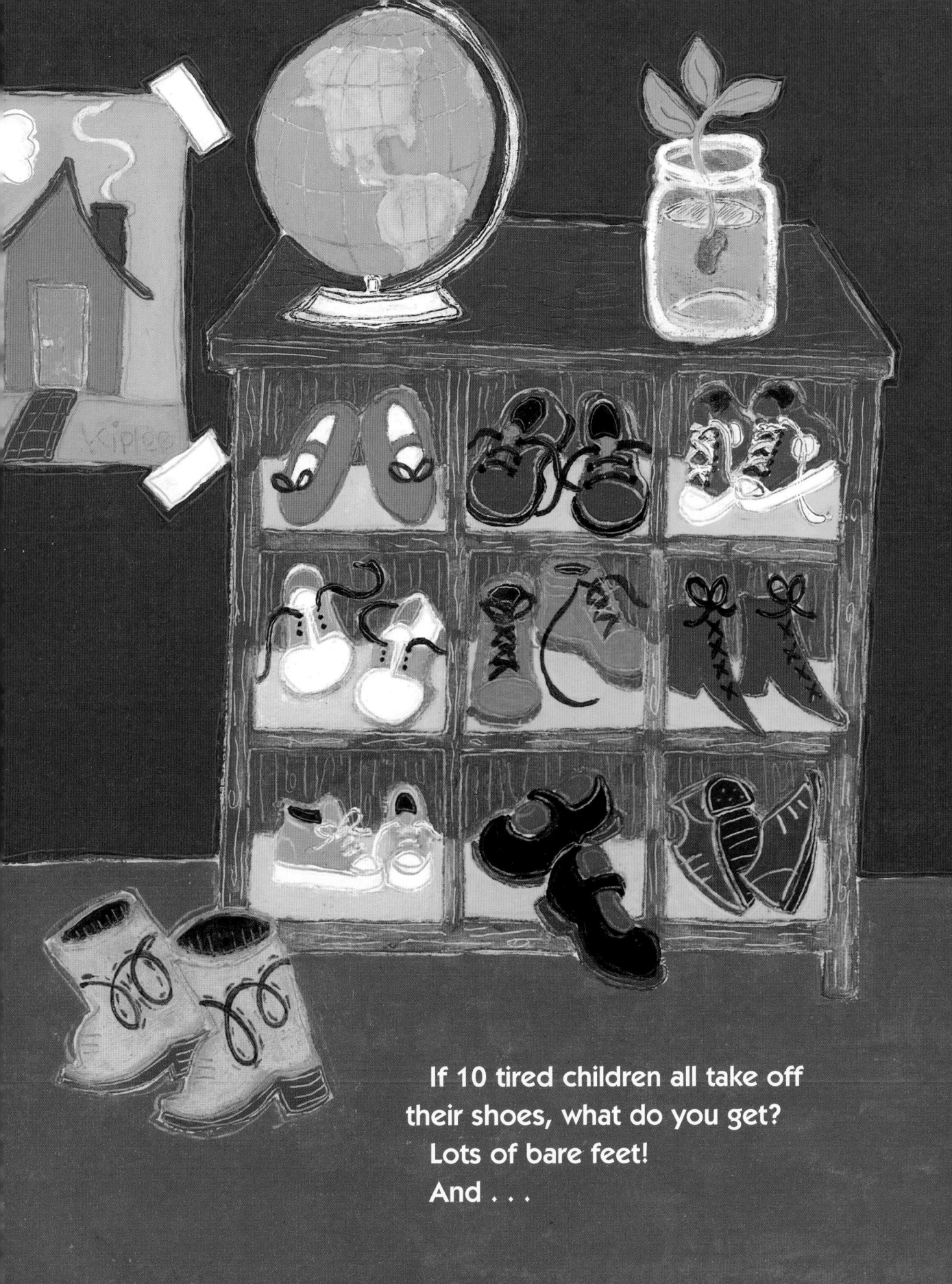

If 10 tired children all take off
their shoes, what do you get?
Lots of bare feet!
And . . .

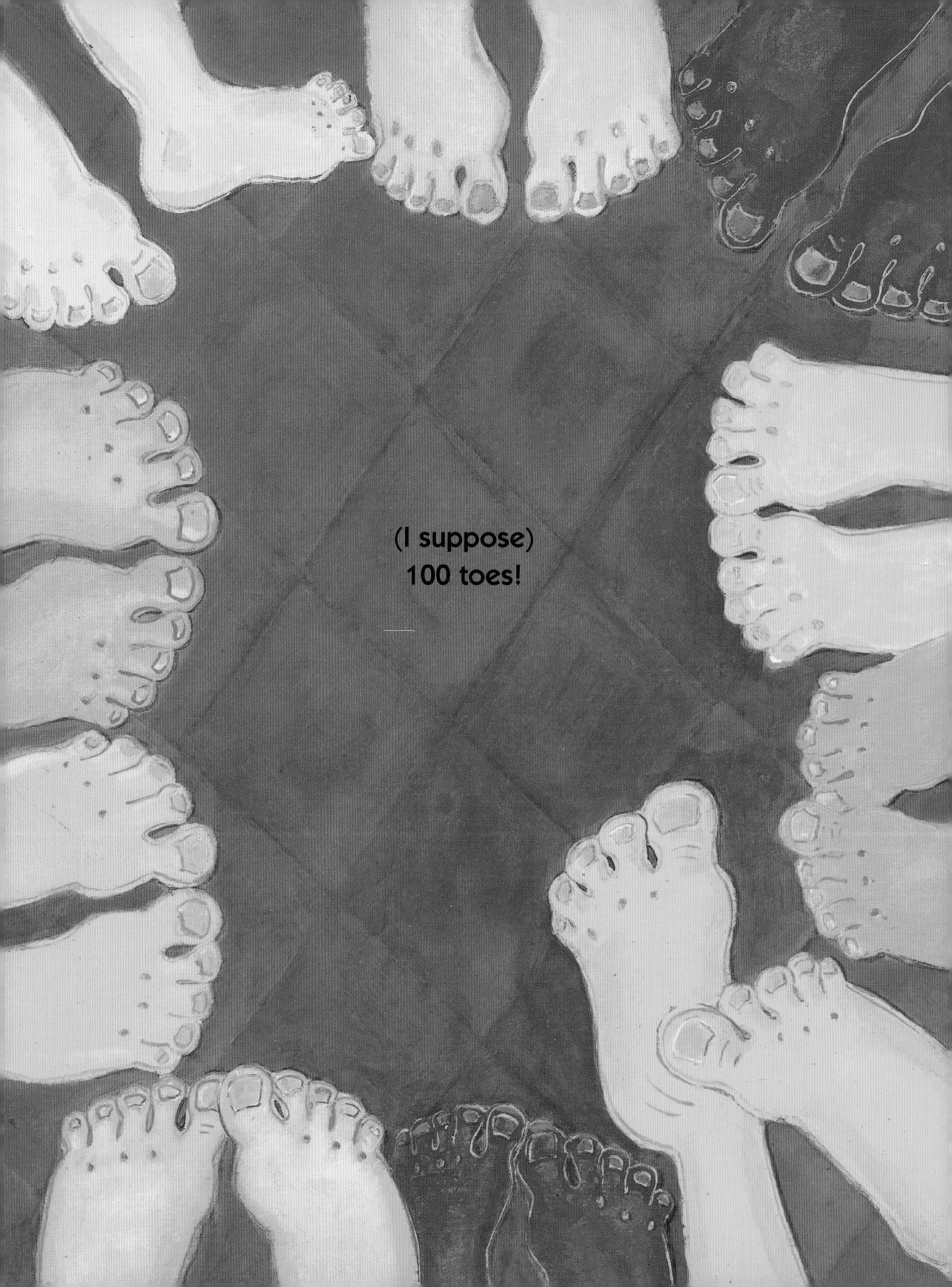

(I suppose)
100 toes!

If you find a tiny bug with
50 legs on one side and 50 on
the other, what do you get?
100 legs.
And . . .

(yes, indeed)
a centipede.

If 20 children each
drop 5 papers on the
floor, what do you get?
100 papers.
And . . .

(I would guess)
an awful mess.

If you eat 10 salty peanuts every minute
for 10 minutes, what do you get?
100 peanuts.
And . . .

(big mistake)
a tummy ache!

If 99 dots are on a clown's suit, what do you get?

100 polka dots. Those . . .

(on his clothes)
plus 1 on his nose.

If 25 bees fly out of a hive, then 25 more and
25 more and 25 MORE, what do you get?
100 bees.
And . . .

(no surprise)
some exercise!

If you put 10 candles on a birthday cake,
and then add 90 more, what do you get?
100 candles.
And . . .

If every day you save
1 penny for 100 days,
what do you get?
100 pennies!
Or . . .

If you pick 75 blackberries,
and your mom picks 25 more,
what do you get?
100 berries.
And . . .

If a train goes by with 99 cars and
then 1 red caboose, what do you get?
100 cars.
And . . .